BUILDING BLOCKS OF CHEMISTRY

CHEMISTRY EVERYWHERE!

Written by William D. Adams

Illustrated by Maxine Lee-Mackie

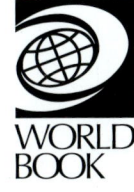

a Scott Fetzer company
Chicago

World Book, Inc.
180 North LaSalle Street
Suite 900
Chicago, Illinois 60601
USA

For information about other World Book publications, visit our website at **www.worldbook.com** or call **1-800-WORLDBK (967-5325).**
For information about sales to schools and libraries, call 1-800-975-3250 (United States),
or 1-800-837-5365 (Canada).

© 2023 World Book, Inc. All rights reserved. This volume may not be reproduced in whole or in part in any form without prior written permission from the publisher.

WORLD BOOK and the GLOBE DEVICE are registered trademarks or trademarks of World Book, Inc.

Library of Congress Cataloging-in-Publication Data for this volume has been applied for.

Building Blocks of Chemistry
ISBN: 978-0-7166-4371-5 (set, hc.)

Chemistry Everywhere!
ISBN: 978-0-7166-4377-7 (hc.)

Also available as:
ISBN: 978-0-7166-4387-6 (e-book)

Printed in India by Thomson Press (India) Limited, Uttar Pradesh, India
1st printing June 2022

WORLD BOOK STAFF
Executive Committee
President: Geoff Broderick
Vice President, Editorial: Tom Evans
Vice President, Finance: Donald D. Keller
Vice President, Marketing: Jean Lin
Vice President, International: Eddy Kisman
Vice President, Technology: Jason Dole
Director, Human Resources: Bev Ecker

Editorial
Manager, New Content: Jeff De La Rosa
Associate Manager, New Product:
   Nicholas Kilzer
Proofreader: Nathalie Strassheim

Graphics and Design
Sr. Visual Communications Designer:
   Melanie Bender
Sr. Web Designer/Digital Media Developer:
   Matt Carrington

Acknowledgments:
Writer: William D. Adams
Illustrator: Maxine Lee-Mackie/
   The Bright Agency
Series Advisor: Marjorie Frank
Additional spot art by Samuel Hiti and
   Shutterstock

TABLE OF CONTENTS

Introduction 4
Chemistry in Your Suitcase 7
Chemistry in Cooking 12
Chemistry of Cars................................ 16
Chemistry of Smell and Taste 20
Chemistry in Airplanes 22
Chemistry of Clouds 24
Chemistry of Ice 25
Chemistry in the Ocean 26
Chemistry of Volcanoes 27
Chemistry in Treasure 30
Chemistry of Rescue 33
Conclusion.. 34
Timeline 36
Can You Believe It?! 38
Words to Know 40

There is a glossary on page 40. Terms defined in the glossary are in type **that looks like this** on their first appearance.

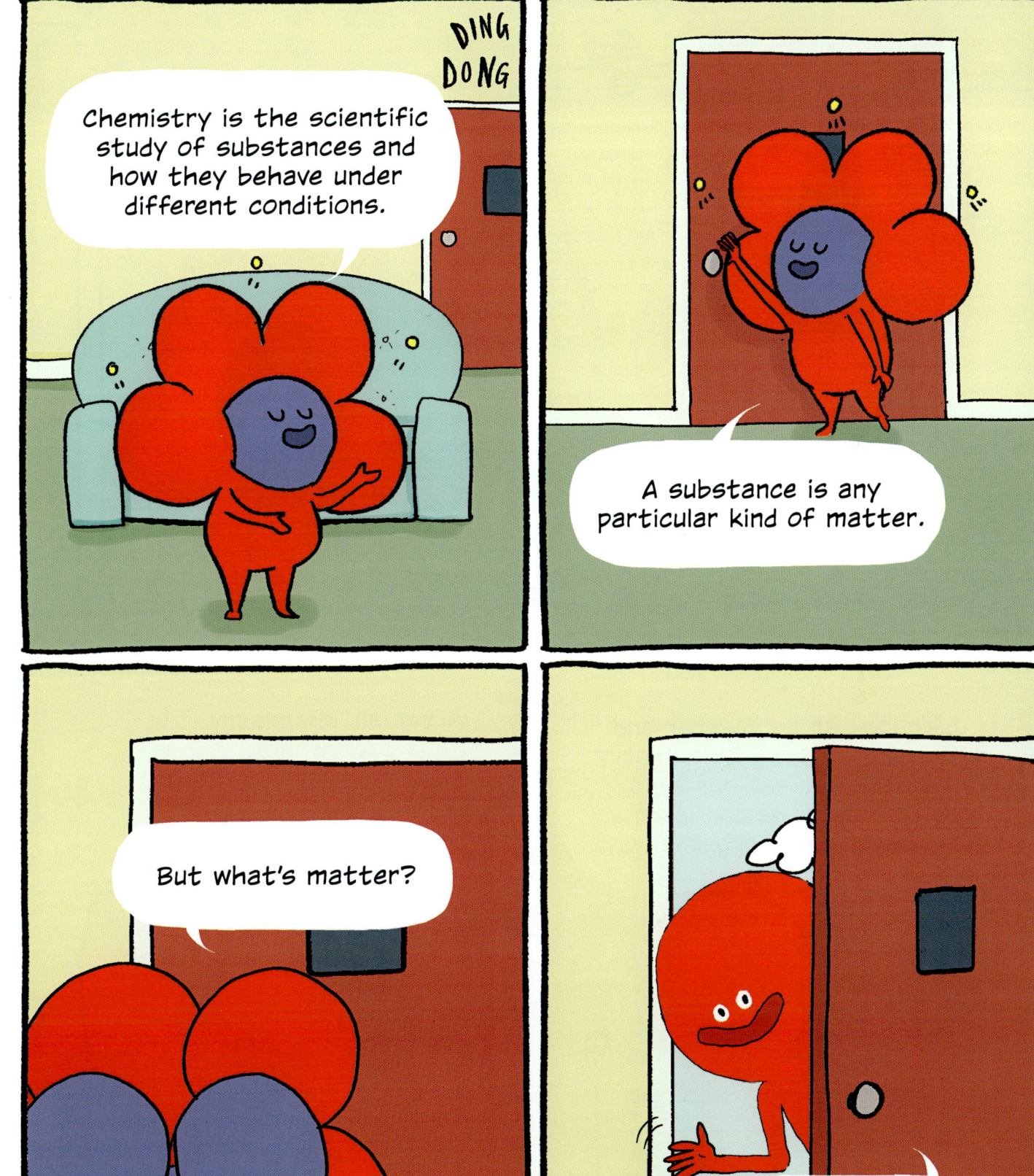

But there are many kinds of light—including kinds you can't see!

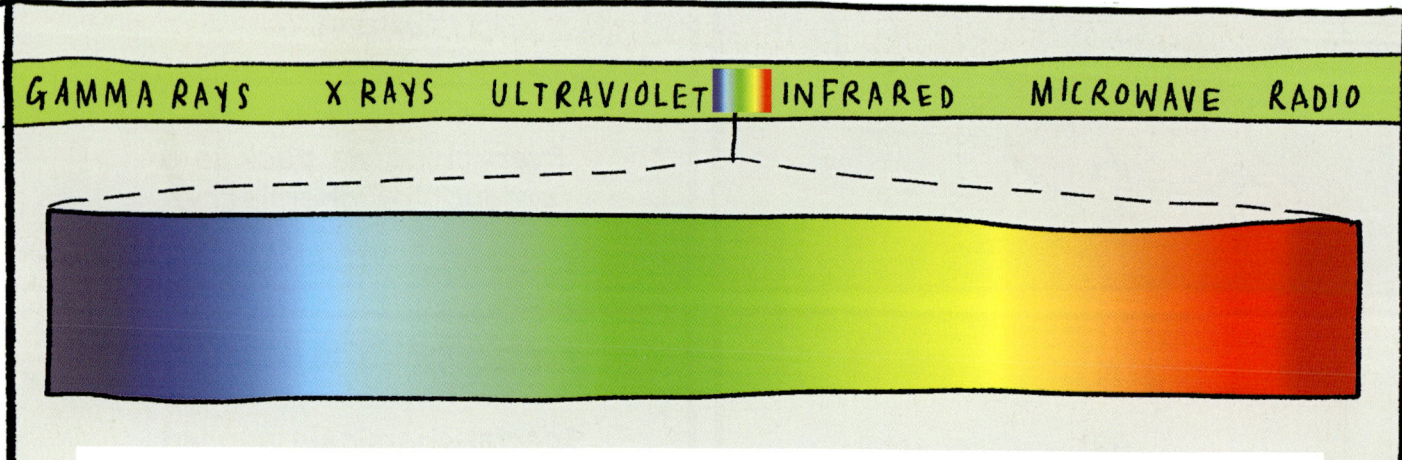

In addition to visible light, the sun puts out invisible, high-energy light that can damage your skin.

Sunscreen contains special **molecules** that block much of this light, keeping your skin healthy and unburned.

If you're going somewhere warm, you're likely to run into biting flies, mosquitoes, ticks, and other bugs that want to feast on you.

The active compounds in insect repellents work by changing how people smell and taste to insects.

Speaking of funk: your skin doesn't smell. But, the billions of tiny bacteria that live on your skin can make you stink. So don't forget to pack–and use–**deodorant!**

Deodorants contain chemical ingredients. Some of these chemicals kill the stinky bacteria. Others prevent the skin from producing the sweat that those bacteria crave.

You'll also want fresh breath and clean teeth while on vacation.

Toothpastes contain chemicals that break down plaque that can form on teeth.

Clothes are made of different fabrics, all of which get their qualities from their chemical composition.

Many fabrics are made of **polymers:** large molecules formed by the bonding of many smaller molecules into a long chain.

Different polymers can give fabrics different properties, making them stretchy, moisture-absorbent, breathable, or itchy.

CHEMISTRY OF CARS

"All cars need energy to start. Powerful car batteries store lots of energy to get the motor running."

"Special chemicals in the battery react to release the energy to start the car."

Cars with gasoline-powered engines harness the power of combustion to drive.

CHEMISTRY IN AIRPLANES

"Oh no—where's the pilot?"

"I can get us there!"

"Airplanes must be light enough to get off the ground but strong enough to survive the stresses of flight."

CLIK

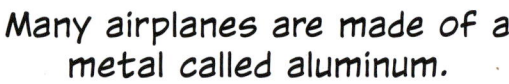
Many airplanes are made of a metal called aluminum.

Aluminum is strong but much lighter than steel.

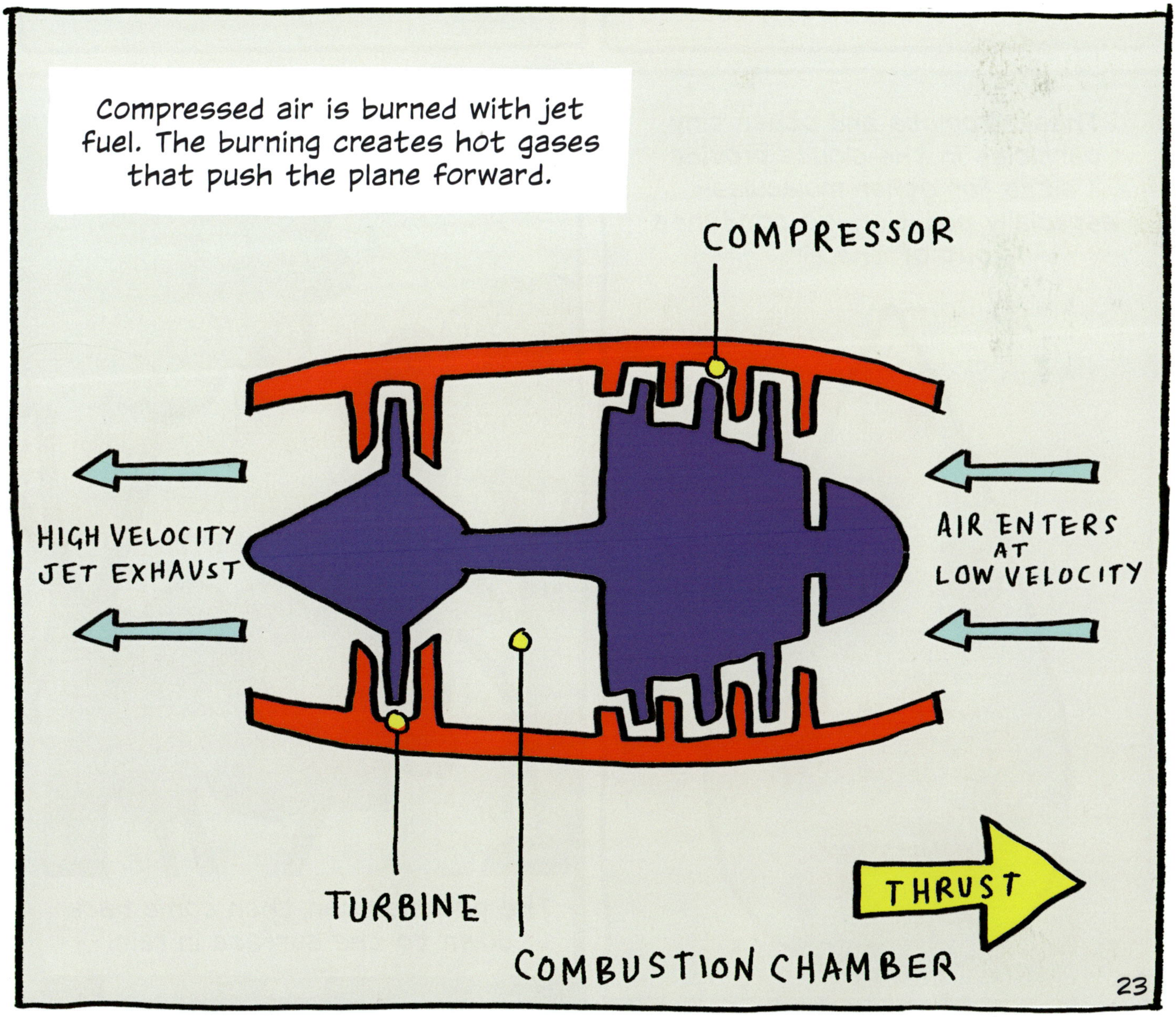

CHEMISTRY OF CLOUDS

There's chemistry in these clouds, too!

Clouds are made of tiny water droplets.

These droplets and other tiny particles in the clouds provide sites for other molecules—especially pollution—to **condense** out of the air.

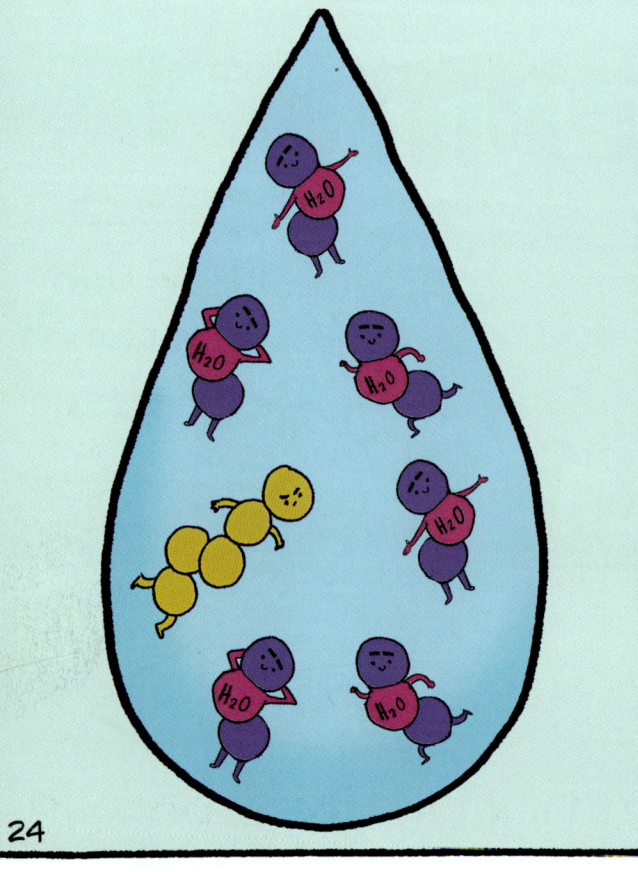

The pollution can then come back down to the surface in rain.

24

CHEMISTRY OF VOLCANOES

Coral reefs often form around volcanic islands. There's chemistry in volcanoes, too!

Volcanoes release the chemicals sulfur dioxide and carbon dioxide into the air when they erupt.

The carbon dioxide they've released over millions of years has helped to trap the sun's heat and keep Earth nice and warm.

But volcanic eruptions can also emit clouds of ash.

The ash is made of tiny pieces of **lava** that solidify in the air. Lava itself is made of several chemical elements.

The ash from huge eruptions can block sunlight, temporarily cooling the planet.

And the ash isn't good for airplane engines!

But volcanic ash provides **minerals** plants use to live and grow.

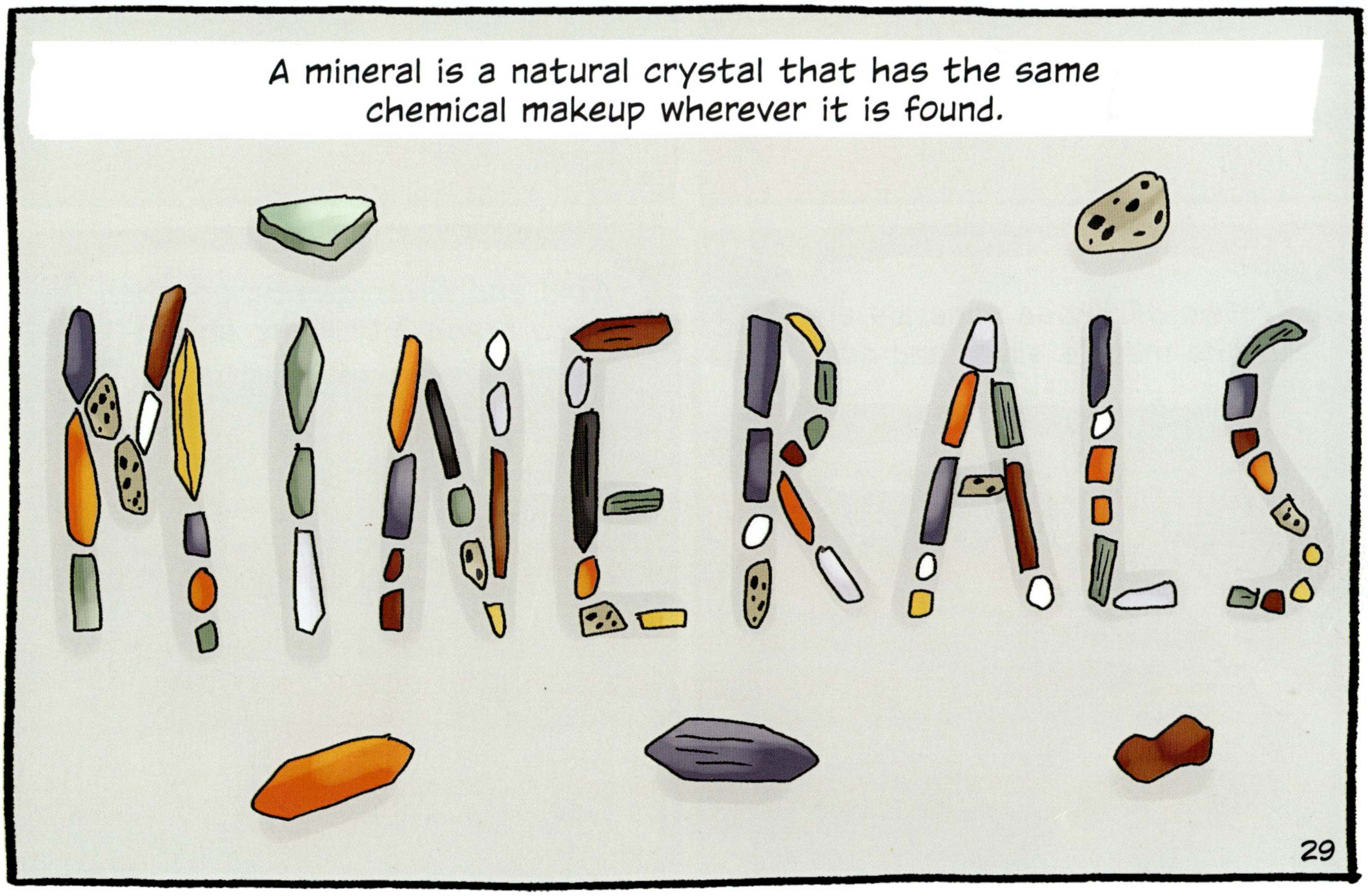

A mineral is a natural crystal that has the same chemical makeup wherever it is found.

MINERALS

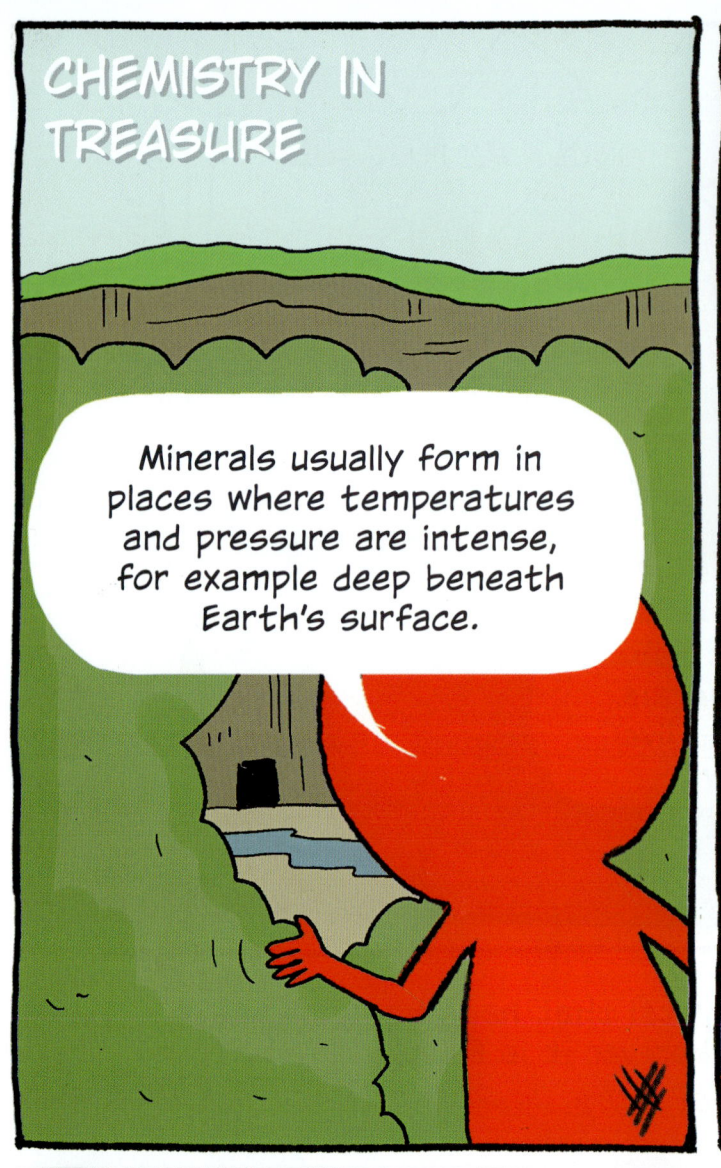

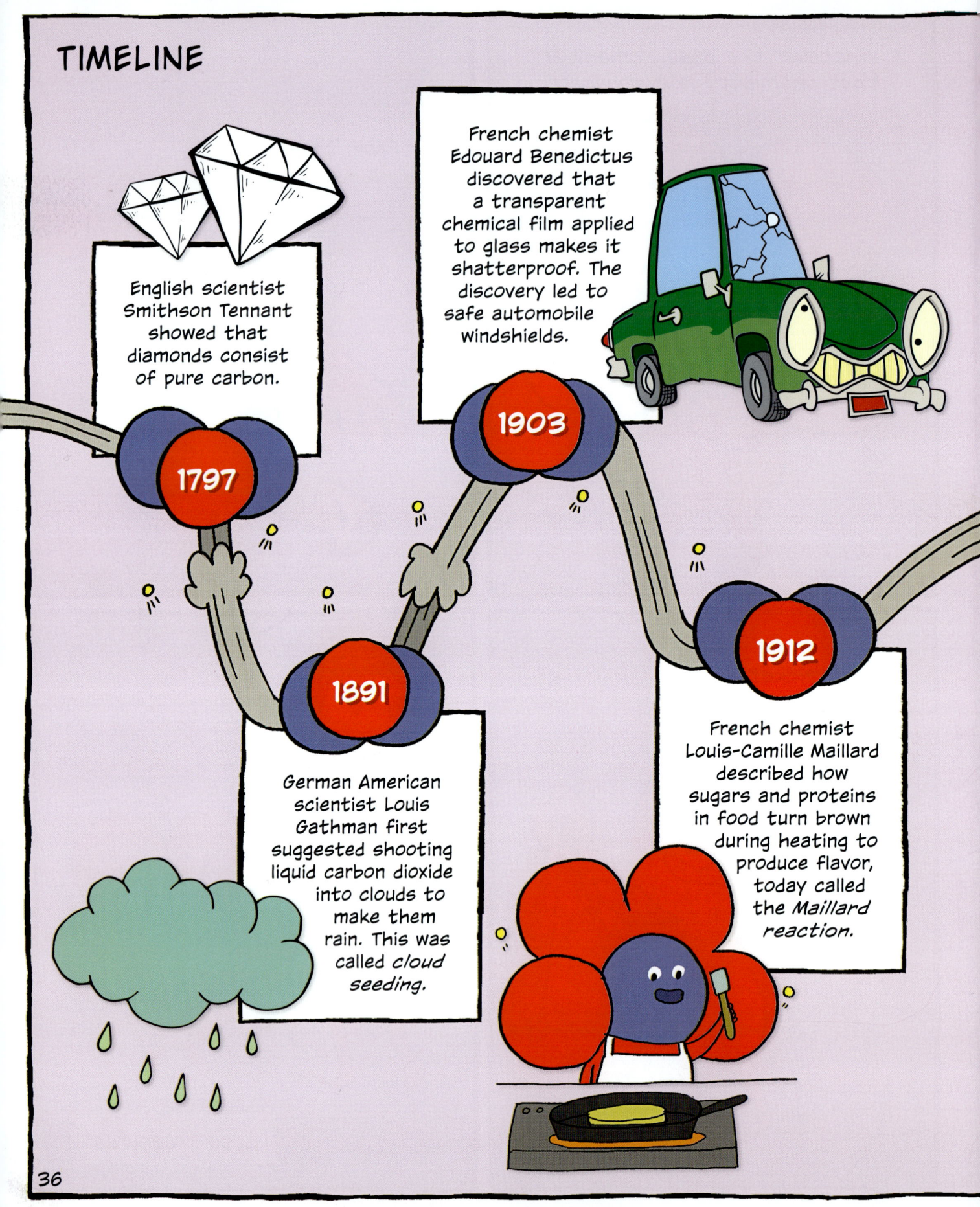

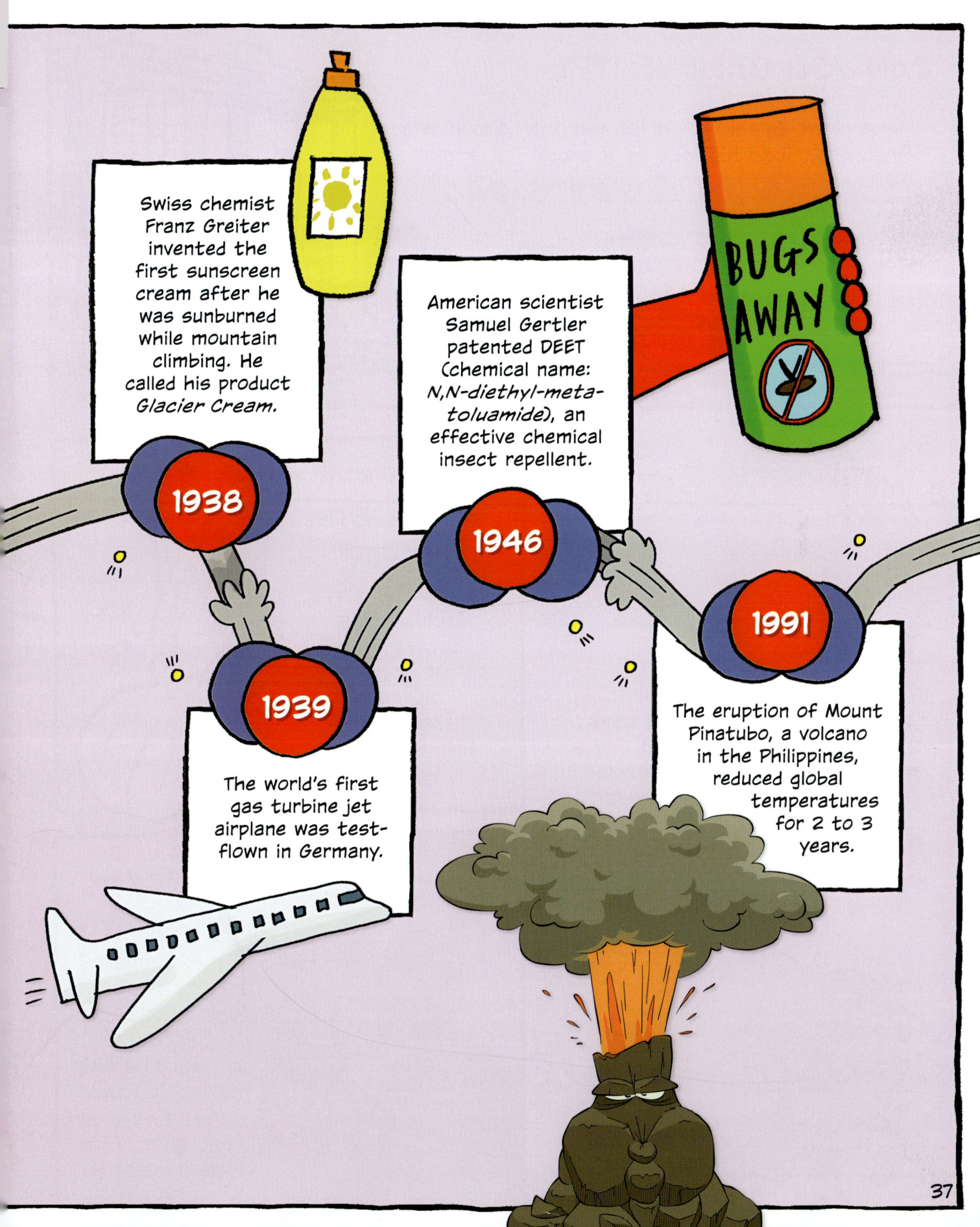

CAN YOU BELIEVE IT?!

The ancient Greek scientist Hero of Alexandria first described his idea of a crude jet engine **about 2,000 years ago!**

The first fire extinguisher, patented in 1723, used a cask of fire-extinguishing liquid and a metal container of gunpowder. Fuses ignited the gunpowder, **which exploded** to scatter the liquid over flames.

The distinctive **smell and flavor** of onions, garlic, and chives comes from sulfur-containing chemical compounds found in these vegetables.

The ancient Egyptians put lumps of perfumed fat on their heads as deodorant. As the fat melted it would cover the skin and give off a pleasing scent.

Red rubies and **blue sapphires** are really the same kind of mineral. They have the same chemical structure. The color difference is due to trace chemical impurities within each gemstone.

Capsaicin is the molecule that gives chili peppers their spicy heat. The molecule interacts with a receptor called TRPV1 on your tongue that sends nerve signals your brain interprets as a **burning sensation.**

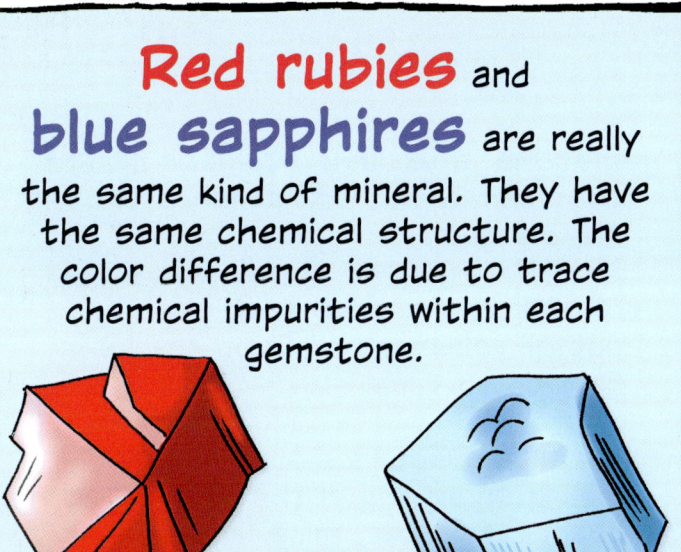

The polymer fabric nylon was used to make the **American flag** placed on the moon by Apollo 11 astronauts in 1969. The fabric's durability has allowed the flag to survive longer on the lunar surface.

About 5,000 different minerals, substances obtained by mining or digging such as quartz and calcite, are known to naturally occur on Earth.

WORDS TO KNOW

atom one of the basic units of matter.

bond the attraction that holds atoms together in groups of two or more.

chemical reaction a process by which one or more substances are chemically converted into one or more different substances.

combustion the act or process of burning.

condense to change from a gas to a liquid.

deodorant any preparation that stops or masks odors, particularly bad odors.

dissolve to break apart a substance, especially by putting it into liquid.

freeze to turn from a liquid into a solid. Water freezes to form hard ice.

gemstone a valuable stone or mineral that can be cut and polished to make a gem.

lava molten (melted) rock that pours out of volcanoes or from cracks in Earth's surface.

mineral a substance obtained by mining or digging in the Earth. Coal, quartz, feldspar, mica, and asphalt are minerals.

molecule two or more atoms chemically bonded together.

polymer a large molecule formed by the chemical linking of many smaller molecules into a long chain.

repellent a chemical substance that repels (drives away) insects or other nuisance animals.

40

+
540 A

Adams, William D. (Childrens
Chemistry everywhere! /
Oak Forest NONFICTION
11/22